The Great Dying

Earth's Most Catastrophic Mass Extinction

Authored by
Zahid Ameer

Published by

Goodword eBooks

DEDICATION

"I dedicate this book to my beloved parents, whose wisdom I hold in the highest regard. Their every word of guidance has been a beacon of light, illuminating the path of my life and shaping the essence of who I am."

The Great Dying

Contents:

The Great Dying

Introduction

Throughout Earth's history, life has undergone periods of incredible growth, diversification, and flourishing, but also catastrophic declines known as **mass extinctions**. These extinctions are not random, isolated events. Rather, they punctuate the evolutionary timeline, each having profound and far-reaching impacts on the planet's biological and ecological framework. Among the five major mass extinction events that have occurred, none is more dramatic and devastating than the **Permian-Triassic extinction event**, also known as **The Great Dying**. Occurring approximately **252 million years ago**, at the close of the Permian period and the dawn of the Triassic, this event forever altered the course of life on Earth.

The Great Dying was not merely a natural disaster; it was a period of extraordinary upheaval and planetary transformation. This cataclysmic event resulted in the extinction of nearly **96% of marine species** and around **70% of terrestrial species**. The sheer scale of the extinction made it the most severe biological crisis in Earth's history, marking the end of the Permian period—an era that had lasted over 47 million years and seen the rise of complex ecosystems both on land and in the oceans. Life in the Permian was diverse and thriving, with ecosystems that included forests, vast coral reefs, and a wide array of invertebrates, amphibians, and reptiles. By the end of this extinction, however, the majority of this diversity had vanished.

The collapse of life during the Great Dying did not occur in a single instant. Rather, it unfolded over hundreds of thousands to millions of years, as Earth underwent a cascade of environmental changes. These changes included severe global warming, ocean acidification, and widespread oxygen depletion—each a consequence of volcanic activity, climatic shifts, and potentially even extraterrestrial factors. The event was not just an ecological turning point but a geological one as well, as it reshaped the physical characteristics of the planet, leading to massive shifts in continental arrangements, sea levels, and atmospheric conditions.

The Scale of the Extinction

To comprehend the magnitude of the Great Dying, it is essential to understand just how extensive the loss of life was. Prior to the extinction, marine ecosystems were extraordinarily rich. Life in the oceans included **trilobites, ammonites, brachiopods**, and vast coral reefs, which played crucial roles in the marine food web. The terrestrial world was equally diverse, populated by **synapsids** (the forerunners of mammals), large amphibians, early reptiles, and dense forests. The extinction, however, brought this flourishing Permian biodiversity to a catastrophic halt.

In the marine realm, up to **96% of species** were wiped out. Entire groups of animals disappeared forever, including certain classes of **ammonoids, echinoderms, and trilobites**. Coral reefs, once expansive and teeming with life, collapsed completely, leaving behind barren underwater landscapes that would not fully recover for millions of years. In the terrestrial realm, the loss of **70% of land species** was equally devastating, as entire ecosystems disintegrated. Massive forest die-offs led to widespread desertification, and many of the dominant species of the time, including the **large amphibians** and **synapsids**, were reduced to just a few surviving lineages.

The enormity of this extinction far surpasses other mass extinction events, including the well-known **Cretaceous-Paleogene extinction**, which wiped out the dinosaurs 66 million years ago. While the Cretaceous event eliminated around **75% of species**, the Great Dying was far more destructive, affecting nearly every ecosystem and life form on Earth. The biosphere was left in ruin, and the recovery of life after this event was staggeringly slow. Some estimates suggest that it took up to **10 million years** for biodiversity to fully rebound, a testament to the severity of the conditions that followed the extinction.

Unraveling the Causes

Understanding the causes behind the Great Dying is one of the most complex and ongoing challenges in paleontology and Earth sciences. Scientists have pieced together a number of factors, all of which likely played a role in creating the perfect storm of environmental devastation. At the heart of the extinction is thought to be the **Siberian Traps**, a massive region of volcanic activity in what is now Russia. These enormous volcanic eruptions, which occurred over a period of approximately **1 million years**, released vast quantities of **carbon dioxide (CO2)** and **sulfur dioxide (SO2)** into the atmosphere. The impact of these gases was catastrophic.

The influx of carbon dioxide into the atmosphere led to an extreme **greenhouse effect**, causing global temperatures to spike dramatically. Some estimates suggest that Earth's average temperature increased by as much as **10–15°C**. This rapid warming triggered a series of secondary effects, such as the **acidification of oceans**, which proved deadly to marine life. Coral reefs, shelled organisms like **brachiopods and ammonites**, and other marine species were unable to adapt to the changing chemistry of the oceans, leading to their extinction.

Compounding the crisis was the depletion of oxygen in the world's oceans, a condition known as **anoxia**. As oceans warmed and the circulation of deep, oxygen-rich waters was disrupted, large swathes of the ocean became oxygen-starved. This lack of oxygen suffocated marine species that relied on it for survival, particularly in the deep seas. Anoxic waters also facilitated the growth of **sulfur-reducing bacteria**, which produced deadly **hydrogen sulfide**—a toxic gas that further poisoned marine ecosystems and even contributed to atmospheric pollution, exacerbating terrestrial extinction.

Another possible contributor to the Great Dying was the release of **methane hydrates** from oceanic sediments. As the planet warmed, these frozen stores of methane—one of the most potent greenhouse gases—may have been destabilized, releasing vast amounts of methane into the atmosphere and accelerating global warming. This so-called "methane pulse" would have further driven temperatures higher, creating a feedback loop of climate instability that pushed many species to the brink of extinction.

Some scientists also speculate that additional factors, such as an **asteroid impact**, similar to the one that ended the reign of the dinosaurs, may have contributed to the Great Dying. However, unlike the well-documented evidence of an impact at the Cretaceous-Paleogene boundary, the evidence for an asteroid or comet impact at the Permian-

Triassic boundary is less clear and remains a subject of debate.

The Aftermath and Evolutionary Repercussions

The aftermath of the Great Dying left Earth almost unrecognizable. With so much life obliterated, ecosystems collapsed, and the biosphere was left in a state of near-total devastation. On land, forests vanished, replaced by vast stretches of arid deserts. In the oceans, biodiversity plummeted, leaving once-thriving reefs and ecosystems devoid of life. The recovery of life was slow and painful, taking millions of years for new species to emerge and fill the ecological void left by the extinction.

However, as with all mass extinctions, the Great Dying also paved the way for new evolutionary opportunities. The species that managed to survive—both in the oceans and on land—would become the ancestors of the new life forms that would dominate the Triassic and beyond. Most notably, the extinction cleared the way for the **rise of the dinosaurs** and the early ancestors of mammals. These new species would come to dominate Earth during the **Mesozoic Era**, reshaping ecosystems in the process.

The Great Dying serves as a critical reminder of the fragility of life on Earth and the delicate balance that sustains it. The environmental crises that led to this mass extinction were driven by forces that we are only now beginning to fully understand, and they offer valuable lessons about the potential consequences of **climate change** and **ecosystem disruption**. As we face our own environmental challenges in the modern world, the Great Dying offers a stark warning about the potential for widespread ecological collapse and the long-term impacts such events can have on the trajectory of life on Earth.

In exploring the causes and consequences of the Permian-Triassic extinction, we gain not only a deeper understanding of Earth's geological and biological history but also critical insights into the resilience of life and the delicate interplay between the planet's systems. Understanding the Great Dying provides us with a window into a past catastrophe and a clearer perspective on how future events might unfold in the face of ongoing environmental changes.

Chapter 1: Life Before the Great Dying

The **Permian period** (298 to 252 million years ago) was a time of remarkable biodiversity, showcasing ecosystems that were rich, complex, and thriving across both land and sea. As the final period of the Paleozoic Era, it was marked by extensive evolutionary development. This period witnessed the flourishing of marine life, the expansion of complex terrestrial ecosystems, and the rise of significant life forms that would later give birth to modern mammals, reptiles, and amphibians. However, the thriving life of this era was on the precipice of an unprecedented catastrophe—the **Great Dying**, which would irreversibly alter the course of Earth's biological history.

1.1 Marine Life in the Permian

The seas during the Permian period were some of the most vibrant and diverse ecosystems in Earth's history. The oceans teemed with life, ranging from small invertebrates to massive, formidable predators. Several dominant groups contributed to the ecological balance of marine environments, but many of these organisms would meet their end in the mass extinction that followed.

Trilobites

Among the most iconic marine creatures of the Paleozoic Era, **trilobites** were a group of extinct marine arthropods that had thrived for nearly **300 million years**. Their distinctive segmented bodies and hard exoskeletons made them successful across various environments. Trilobites had evolved into an astonishing range of species, inhabiting many ecological niches. Some were predators, others scavengers, while some were filter feeders. Trilobites had sophisticated compound eyes, which gave them excellent vision, a crucial advantage in the complex and competitive marine ecosystems.

Despite their long success, trilobites had already experienced significant declines in diversity by the end of the Permian period. A combination of climate change and competition from other marine organisms had gradually reduced their numbers. Nevertheless, some species still thrived in the shallow Permian seas—until the **Great Dying** wiped out the last remaining trilobites, marking the end of one of Earth's most iconic animal groups.

Brachiopods and Ammonites

Other dominant marine species during the Permian period included **brachiopods** and **ammonites**, both of which played crucial roles in the marine food web.

Brachiopods, often confused with bivalves like clams, were shelled invertebrates that anchored themselves to the sea floor, using a fleshy stalk known as a **pedicle**. With over 30,000 fossil species identified, brachiopods had achieved remarkable evolutionary success by the Permian. They fed by filtering nutrients from the water, playing a vital role in sustaining the marine ecosystems of the time. However, their stationary lifestyle made them particularly vulnerable to environmental changes. While some brachiopod species survived the Great Dying, many did not, and their dominance in the oceans waned significantly after the event.

Ammonites, related to modern-day squids and octopuses, were a group of marine mollusks that developed spiral-shaped shells. Highly adaptive and mobile, ammonites were the **apex predators** of their time, hunting smaller marine organisms with their tentacles. Their buoyant shells allowed them to navigate various ocean depths, giving them a significant evolutionary advantage. Despite their resilience and widespread presence, most ammonite species were unable to survive the drastic environmental changes of the Permian-Triassic extinction. While some ammonites re-emerged in the Triassic and flourished in later periods, their diversity was never the same after the Great Dying.

Coral Reefs

One of the most striking features of the Permian seas was the presence of vast **coral reefs**, which, like today's reefs, provided critical habitats for a wide variety of marine life. These reefs were primarily built by **rugose corals** and **tabulate corals**, which are now extinct but were essential in creating the complex, three-dimensional structures of Permian marine ecosystems. The coral reefs were home to numerous species of fish, mollusks, sponges, and echinoderms, and they played a vital role in maintaining marine biodiversity by offering shelter and feeding grounds.

However, coral reefs are extremely sensitive to changes in temperature and water chemistry. During the end of the Permian, the dramatic rise in atmospheric carbon dioxide levels and the corresponding increase in ocean acidity had a devastating effect on coral reefs. Many of these ancient reef ecosystems collapsed entirely during the Great Dying, contributing to the overall collapse of marine biodiversity.

1.2 Terrestrial Life

While the Permian seas were teeming with life, the terrestrial ecosystems were no less diverse and dynamic. The continents had merged into a supercontinent known as **Pangaea**, creating vast stretches of land with varied climates and environments. This enabled the proliferation of diverse ecosystems, from swampy wetlands to arid deserts. On land, life was dominated by various species of **amphibians**, **reptiles**, and **synapsids**—the latter being the ancestors of modern mammals.

Synapsids

Among the most significant groups of terrestrial vertebrates in the Permian were the **synapsids**, often referred to as **mammal-like reptiles**, although they were not true reptiles. Synapsids were characterized by a single opening in the skull behind the eyes (the **temporal fenestra**), which allowed for larger jaw muscles and more powerful bites. This group included a wide range of species, from small, insect-eating creatures to large, powerful predators.

One of the most famous synapsids of the Permian was the **Dimetrodon**, a predatory species often mistakenly associated with dinosaurs. The Dimetrodon was distinguished by its **large sail-like structure** on its back, which may have been used for regulating body temperature or for display during mating rituals. As an apex predator, Dimetrodon dominated its environment, preying on smaller reptiles and amphibians. Despite its evolutionary success, like many other synapsids, it would be largely wiped out during the Great Dying.

Synapsids played an important role in the evolution of mammals, and while many species perished in the Permian extinction, a few managed to survive and eventually gave rise to the true mammals of the Mesozoic and beyond.

Large Amphibians

The Permian period also saw the dominance of **large amphibians**, particularly in the wetland and swampy environments that were common during this era. These amphibians, often much larger than their modern descendants, were major predators in freshwater environments and relied on water for reproduction, much like modern amphibians.

Temnospondyls, for example, were a group of large, often crocodile-like amphibians that flourished during the Permian. With strong, sharp teeth and powerful limbs, they were formidable predators in their aquatic habitats. However, the widespread drying of the climate during the Permian, coupled with the dramatic environmental changes at the end of the period, led to the decline of many amphibian species. The survivors would evolve into the frogs, salamanders, and newts we see today.

Pelycosaurs

Pelycosaurs were another dominant group during the early and middle parts of the Permian. These early reptiles were among the largest land animals of their time, occupying various ecological niches, from herbivores to carnivores.

While pelycosaurs are often grouped with synapsids, they are considered more primitive and less mammal-like. Like the synapsids, pelycosaurs displayed a wide range of body forms, with some species resembling modern reptiles and others developing the distinctive back sails similar to Dimetrodon.

As top predators and dominant herbivores, pelycosaurs shaped the terrestrial ecosystems of the Permian. However, by the end of the period, most pelycosaurs had gone extinct, replaced by more advanced forms of synapsids and reptiles.

The Complex Precursor to Collapse

The **late Permian ecosystems** were rich in complexity, with intricate food webs, diverse species, and widespread ecological interactions. Marine life thrived in the warm, shallow seas, while terrestrial life evolved to occupy nearly every available niche on land. Yet, this remarkable period of biological diversity and adaptation was about to face one of the most dramatic and catastrophic events in Earth's history.

In the blink of a geological eye, the balance of life on Earth would be shattered. Rising temperatures, ocean acidification, volcanic activity, and a host of other factors would converge to initiate a mass extinction so severe that nearly all life on the planet would be affected. The world that emerged after the **Great Dying** was a far cry from the one that existed before, with many of the dominant species lost forever, and new forms of life gradually emerging to take their place.

This period of life before the Permian-Triassic extinction gives us a glimpse into a time when Earth's ecosystems were at their peak, only to be brought to the brink of total collapse by forces beyond their control.

Chapter 2: The Cataclysm Unfolds

The Permian-Triassic extinction, or the Great Dying, was not a sudden catastrophe that wiped out life overnight. Instead, it was a complex, multifaceted series of events that unfolded over hundreds of thousands, if not millions, of years. This mass extinction resulted from a confluence of environmental crises that gradually eroded Earth's ecosystems. At the core of these crises were volcanic activity, climate change, oceanic anoxia (oxygen depletion), and the release of deadly gases like methane. Each of these factors compounded the others, creating a cascading environmental collapse that suffocated life on land and in the oceans.

2.1 The Siberian Traps: The Culprit of Doom

One of the most significant culprits behind the Great Dying is the **Siberian Traps**, an enormous expanse of volcanic rock in what is now northern Russia. The Siberian Traps represent one of the largest volcanic provinces in Earth's history, covering an area roughly the size of the continental United States. The volcanic activity in this region unleashed a flood of **basalt lava flows**, covering large parts of the planet and releasing an enormous volume of gases that altered the atmosphere and oceans.

The volcanic eruptions of the Siberian Traps lasted for about a million years, and their effects were devastating. Here are the major ways these eruptions contributed to the Great Dying:

2.1.1 Massive Release of Carbon Dioxide (CO_2) and Sulfur Dioxide (SO_2)

The Siberian Traps eruptions emitted vast amounts of **carbon dioxide (CO_2)** and **sulfur dioxide (SO_2)** into the atmosphere. The release of these gases had a two-fold effect:

- **Greenhouse Warming**: The large amounts of CO_2 pumped into the atmosphere created an intense **greenhouse effect**, where heat from the sun was trapped within the Earth's atmosphere. This caused global temperatures to rise sharply, potentially by **10-15°C**. The warming led to changes in ocean circulation and significantly disrupted weather patterns, contributing to the breakdown of ecosystems that had evolved in a more stable Permian climate.
- **Acid Rain**: The sulfur dioxide released by the eruptions likely combined with water vapor in the atmosphere, forming sulfuric acid. This led to **acid rain**, which would have devastated plant life on land

and contributed to **ocean acidification**, further exacerbating the collapse of marine ecosystems.

2.1.2 A Runaway Greenhouse Effect

As global temperatures increased, a dangerous feedback loop known as the **runaway greenhouse effect** may have been triggered. This is a phenomenon where initial warming causes conditions that lead to further, uncontrollable warming. For example:

- Higher temperatures cause more water vapor to enter the atmosphere, which traps more heat because water vapor is itself a potent greenhouse gas.
- Warmer temperatures melt ice and snow, reducing the Earth's albedo (reflectivity), meaning less sunlight is reflected back into space, and more is absorbed, further heating the planet.

The resulting **climate instability** wreaked havoc on both terrestrial and marine environments. Plants, which are highly sensitive to changes in temperature and atmospheric chemistry, began to die off in large numbers, causing a collapse in food webs. With food sources dwindling, herbivores began to starve, and predators, in turn, faced extinction.

2.1.3 Ocean Acidification

Perhaps the most lethal consequence of the Siberian Traps eruptions was their effect on the oceans. The high levels of CO_2 not only warmed the planet but also dissolved in ocean water, creating **carbonic acid**. This process led to **ocean acidification**, a dramatic change in the ocean's chemistry that had catastrophic consequences for marine life. Marine organisms, particularly those with calcium carbonate shells or skeletons—such as **corals, ammonites, and brachiopods**—struggled to survive as acidic waters dissolved their protective structures.

This acidification also affected other forms of marine life, leading to the collapse of the marine food chain. Coral reefs, which serve as critical habitats for many species, were decimated. Without these structures, the biodiversity of marine ecosystems plummeted.

2.1.4 Release of Toxic Gases: Methane Amplifies Instability

As if the release of carbon dioxide and sulfur dioxide were not enough, the Siberian Traps eruptions may have triggered the release of another deadly gas—**methane**. Methane is a highly potent greenhouse gas, much more effective at trapping heat than CO_2. The warming oceans likely destabilized methane hydrates—crystalline structures found in ocean sediments that trap methane molecules within them. When ocean temperatures rise, these hydrates can break down, releasing large quantities of methane into the atmosphere.

The methane released during the Siberian Traps eruptions would have created an even more severe greenhouse effect, intensifying global warming and contributing to the collapse of ecosystems on land and at sea.

2.2 Climate Change and Oceanic Anoxia

One of the most deadly consequences of the global warming triggered by the Siberian Traps eruptions was the onset of **oceanic anoxia**—a condition in which the oceans become depleted of oxygen. The warming of the planet caused ocean temperatures to rise, which disrupted ocean circulation patterns. This disruption prevented oxygen from being transported to deeper ocean layers, leading to **widespread anoxia**.

2.2.1 The Collapse of Ocean Circulation

Under normal conditions, cold, oxygen-rich waters from the surface sink to the depths of the ocean, providing oxygen to marine organisms living at all levels. However, as global temperatures soared during the Great Dying, the oceans became **stratified**—with warm water at the surface and cooler water trapped below. This prevented oxygen from reaching deeper parts of the ocean.

Without oxygen, marine organisms that depended on it for survival began to die off in massive numbers. Coral reefs, already struggling from acidification, faced an additional threat as oxygen levels plummeted. **Anoxic conditions** became widespread, turning large portions of the ocean into "dead zones" where little to no life could survive.

2.2.2 Marine Extinctions

The anoxic oceans proved to be a death sentence for many marine species. Fish, mollusks, and other marine animals died en masse, as the lack of oxygen suffocated them. Deep-sea environments, which rely on sinking organic material for food, were particularly affected, as this food source dwindled with the collapse of the surface ecosystems.

In addition to marine animals, **plankton**, which form the base of the marine food chain, also suffered. Plankton are vital for producing oxygen and supporting marine biodiversity. Their decline had far-reaching consequences, further destabilizing the already fragile ecosystems.

2.3 Methane Release and the "Death Pulse"

The warming oceans not only led to anoxia but also contributed to the release of **methane** stored in oceanic sediments. This release occurred in a phenomenon known as the **"methane death pulse,"** which refers to the sudden and catastrophic release of methane into the atmosphere.

2.3.1 Methane Clathrates and Warming Oceans

Methane clathrates are icy structures found under the seabed that trap methane within their lattices. These structures are stable under high pressure and low temperatures, but when ocean temperatures rise, they can destabilize, releasing their methane into the water and eventually into the atmosphere.

During the Great Dying, as the oceans warmed from volcanic activity, methane clathrates began to break apart, releasing large amounts of methane into the atmosphere. Methane is a **potent greenhouse gas**—much more effective at trapping heat than carbon dioxide—so its release would have significantly accelerated global warming.

2.3.2 The Methane Pulse and Ecosystem Collapse

The release of methane in the so-called **methane pulse** further destabilized Earth's climate. The added methane created a rapid and intense spike in global temperatures, amplifying the effects of the initial volcanic warming. This led to what is often referred to as a **"runaway greenhouse effect,"** where the warming triggered by methane release caused further destabilization of methane hydrates, resulting in even more methane entering the atmosphere.

This runaway greenhouse effect had catastrophic consequences for life on Earth. Land animals, already struggling with changing temperatures and dwindling food sources, faced increasingly hostile conditions. Marine life, too, continued to suffer from ocean acidification, anoxia, and now heightened temperatures from the methane pulse. Entire ecosystems collapsed under the combined weight of these environmental disasters.

Conclusion

The Great Dying was a result of complex and interrelated environmental factors, primarily driven by the volcanic eruptions of the Siberian Traps. The massive release of carbon dioxide, sulfur dioxide, and methane led to a **global climate crisis**, warming the planet, acidifying the oceans, and depleting oxygen from the seas. This series of events unleashed a cascade of extinctions, wiping out nearly all marine life and the majority of terrestrial species. The **oceanic anoxia**, **methane pulse**, and runaway greenhouse effect created an inhospitable world, from which life would take millions of years to recover. The lessons learned from this extinction event offer a stark reminder of the fragility of Earth's ecosystems and the profound consequences of atmospheric and climate shifts.

Chapter 3: The Consequences of the Great Dying

The sheer scale of the **Permian-Triassic extinction event**, or **The Great Dying**, transformed Earth's biosphere on an unprecedented level. The planet, once teeming with life in the seas, on land, and in the skies, became a barren wasteland, almost devoid of the rich ecosystems that had flourished for millions of years. This extinction, which eradicated up to **96% of marine species** and **70% of terrestrial species**, reshaped the biological landscape in ways that still echo through time. The aftermath left a planet struggling to recover, with massive disruptions in both marine and land-based ecosystems. The ripple effects from this extinction lasted for millions of years, fundamentally altering evolutionary pathways and ecological interactions.

3.1 Marine Devastation

Of all the environments affected by the Great Dying, **marine ecosystems** experienced the most profound devastation. The oceans, which had once supported a vast and diverse array of species, were thrown into chaos as oxygen levels plummeted, temperatures soared, and acidic conditions took hold.

Coral Reef Collapse

Before the Great Dying, the seas were adorned with **vibrant coral reefs**, akin to today's coral ecosystems but dominated by ancient organisms. These reefs supported a wide variety of marine life, from small filter feeders to larger predators. However, during the extinction event, **ocean acidification** and rising temperatures spelled doom for these reef systems. The high levels of **CO2** released from volcanic activity in the **Siberian Traps** not only caused global warming but also dissolved into the oceans, forming **carbonic acid**. The resulting drop in ocean pH disrupted calcium carbonate formation, the building block of coral reefs, making it impossible for coral organisms to maintain their skeletons.

With the collapse of coral reefs, vast marine ecosystems lost their structural backbone. Coral reefs are not only home to countless species but also act as natural barriers, protecting coastlines and providing habitats for both prey and predators. Their collapse left entire marine food webs destabilized, amplifying the effects of extinction across species that relied on them for survival.

Extinction of Key Marine Species

Several iconic marine species groups were completely wiped out or drastically reduced during the Great Dying:

- **Ammonites**, once plentiful and diverse, became nearly extinct, with only a few lineages surviving into the Mesozoic.
- **Trilobites**, which had thrived for over 300 million years, were completely eradicated by this event.
- **Brachiopods**, which had dominated the seafloor for much of the Paleozoic, experienced a near-total collapse.

These species were integral parts of the marine food web. The extinction of filter feeders, such as **brachiopods**, disrupted nutrient cycling, while the loss of predators, such as **ammonites**, threw marine ecosystems out of balance. Species higher up the food chain, like **fish**, suffered significantly as a result.

Fish Populations and the Collapse of the Marine Food Web

The loss of biodiversity at the base of the marine food web created cascading effects for larger organisms. **Fish populations** plummeted as their prey—smaller invertebrates and plankton—dwindled due to poor water conditions, lack of oxygen, and the collapse of coral habitats. Fish that relied on coral reefs for shelter and food were particularly hard hit.

The **anoxic (oxygen-depleted) conditions** in the oceans were perhaps the most lethal factor. As the oceans warmed, the stratification of ocean layers prevented the mixing of oxygen-rich surface waters with deeper waters, creating vast **dead zones**. Marine life suffocated as oxygen levels fell to critical lows, especially affecting larger, more active animals that needed more oxygen to survive.

The collapse of fish populations further unbalanced the marine food web. Without predators to keep them in check, some species overpopulated, while others disappeared entirely. This breakdown left ecosystems in disarray, and the slow recovery of marine biodiversity ensured that for millions of years, oceans remained relatively barren in comparison to their pre-extinction state.

3.2 Land Ecosystem Collapse

The Permian-Triassic extinction was equally catastrophic for terrestrial ecosystems. What were once lush forests, swamps, and verdant landscapes filled with diverse life forms became barren deserts and wastelands. The once-dominant **synapsids** (mammal-like reptiles) and **amphibians** were particularly hard-hit, along with the **giant insects** and plants that had characterized the Permian world.

Collapse of Synapsids and Other Land Vertebrates

Before the Great Dying, **synapsids**, such as the **Dimetrodon** and **Gorgonopsians**, were among the dominant land-based predators. These mammal-like reptiles were highly successful, but their numbers were decimated by the environmental shifts of the extinction event. The combination of **extreme heat, acid rain, and lack of food** decimated both herbivores and carnivores. Herbivorous synapsids, which fed on the now-disappearing plant life, were unable to sustain themselves, and their predators followed suit.

The collapse of plant life further exacerbated the loss of land vertebrates. With **forests dying** off and **vegetation disappearing** due to changing climates, particularly intense droughts, herbivores struggled to find food, leading to mass starvation. Without herbivores to prey upon, carnivores, including synapsids and amphibians, also faced extinction.

Forest Death and Desertification

Perhaps one of the most visible consequences of the Great Dying on land was the collapse of entire **forest ecosystems**. Much like the reefs of the ocean, forests were foundational to life on land, providing food, shelter, and oxygen. However, the intense climate change caused by volcanic gases, global warming, and acid rain led to the large-scale destruction of plant life. **Acid rain**, in particular, damaged soil fertility and the leaves of plants, halting photosynthesis. Coupled with an increase in arid conditions and desertification, forests could no longer survive in many parts of the world.

The loss of these forests had far-reaching consequences. Not only did it lead to the **loss of habitats** for countless species, but it also created vast stretches of desert where life struggled to regain a foothold. The disappearance of forests and the resulting desertification marked a significant ecological shift, with many of Earth's most fertile regions turning into inhospitable wastelands.

Herbivores and Predators: A Chain Reaction

As plant life withered and died, **herbivores**—the primary consumers—were the first to experience the full brunt of the extinction. Species that had thrived on the lush vegetation of the Permian period could no longer find sustenance. As herbivores died off in great numbers, the **carnivores** that preyed upon them began to starve as well.

This led to a devastating **cascading effect** within the terrestrial food chain. With the collapse of both plant life and the animals that relied on it, entire ecosystems crumbled. The interdependence of species became their downfall as the extinction created a domino effect, leading to the eventual collapse of even resilient species.

3.3 Recovery and the Rise of New Life Forms

Despite the unimaginable devastation of the Great Dying, **life did eventually recover**, although it took millions of years for ecosystems to stabilize and diversify once again. The survivors of this event laid the groundwork for a new era of life on Earth, radically different from the Permian world that had preceded it.

Slow Recovery of Ecosystems

The **recovery process** after the Great Dying was painfully slow. Marine ecosystems remained barren for millions of years. In the immediate aftermath, only a few hardy species managed to cling to life in the oxygen-depleted, overheated waters. **Algae** and some simple **invertebrates** were among the few organisms that began to repopulate the oceans, but the complex ecosystems that had once thrived would not return for a long time.

On land, the survivors were primarily **small, burrowing animals** that could withstand harsh conditions. Species that could survive in environments with limited food and water, such as some **reptiles** and early **therapsids** (the ancestors of mammals), managed to endure the hardships of the post-extinction world. However, the landscapes were mostly barren, and the absence of forests and fertile soils made recovery difficult for herbivores and the species that depended on them.

The Mesozoic Era and the Rise of Dinosaurs

Although the Permian world had been dominated by synapsids and amphibians, the survivors of the Great Dying included a new group that would come to dominate the land: the **archosaurs**. These reptiles, which included the ancestors of **crocodiles, pterosaurs**, and **dinosaurs**, were remarkably resilient and well-adapted to the new post-extinction environment. Over the next several million years, **dinosaurs** rose to prominence, eventually becoming the dominant terrestrial animals of the **Mesozoic Era**.

During this period, new ecological niches were filled by species that evolved in the wake of the extinction. While many of the creatures of the Mesozoic bore some resemblance to their Permian ancestors, they represented a distinct evolutionary shift. The rise of **dinosaurs** fundamentally reshaped terrestrial ecosystems, and this new age would last for over 160 million years.

The Evolution of Mammalian Ancestors

While dinosaurs dominated the land, the few surviving **synapsids**, such as the **cynodonts**, gradually evolved into **early mammals**. These small, nocturnal creatures were not yet dominant but were able to carve out a niche in the shadow of the dinosaurs. The extinction had cleared the evolutionary path for new life forms, allowing mammals to eventually rise to prominence millions of years later.

The legacy of the Great Dying is a profound reminder of the **resilience of life** and the **power of evolution** to shape and adapt to a changing world. While the extinction wiped out the majority of species, it also paved the way for the emergence of new life forms, including the ancestors of humans. In this sense, the Great Dying was not only an ending but also a **new beginning** for life on Earth.

Chapter 4: Theories and Debates

The **Permian-Triassic extinction event**, or the **Great Dying**, remains one of the most extensively studied mass extinction events in Earth's history due to its sheer scale and impact. While the **Siberian Traps volcanic eruptions** theory is the leading explanation for this event, the extinction was so complex and multifaceted that additional theories have been proposed. Scientists continue to debate the exact causes, recognizing that the extinction might have been the result of multiple factors interacting in devastating ways. This chapter explores the alternative theories and potential contributing factors that could have played significant roles in the Great Dying, including **asteroid impact**, **marine instabilities**, and the formation of the **supercontinent Pangaea**.

4.1 Asteroid Impact: A Hypothetical Cosmic Trigger

One of the most intriguing theories proposed to explain the Great Dying is the possibility of an **asteroid or comet impact**. This idea parallels the more widely known **Cretaceous-Paleogene (K-Pg) extinction event**, which took place around 66 million years ago and led to the extinction of the dinosaurs. The K-Pg extinction has been strongly linked to a large asteroid impact at **Chicxulub**, near present-day Mexico. Given the success of this impact hypothesis in explaining the dinosaur extinction, some scientists have considered whether a similar event might have triggered the Permian-Triassic extinction.

4.1.1 The Evidence for an Impact

Proponents of the asteroid impact theory argue that the Permian extinction might have been caused or exacerbated by an extraterrestrial impact. This hypothesis suggests that:

- An asteroid or comet impact could have caused massive shockwaves, wildfires, and atmospheric disturbances, which may have contributed to the extreme climate changes during this period.
- The impact could have generated **dust clouds** and **aerosols** that blocked sunlight, leading to global cooling, followed by extreme warming once the particles settled. These abrupt climate shifts would have been detrimental to both marine and terrestrial ecosystems.

However, unlike the K-Pg extinction, direct evidence of such an impact has remained elusive. Specifically, researchers have not found a definitive **impact crater** or a layer of **iridium** (a rare element often associated with asteroid impacts) that could be traced to the Permian-Triassic boundary. Some scientists have argued that the impact, if it occurred, might not have left clear geological traces or that the evidence has been obscured over millions of years due to erosion or tectonic activity.

4.1.2 The Bedout Crater Hypothesis

In the early 2000s, researchers suggested the presence of a possible impact structure called the **Bedout High**, located off the coast of Western Australia. The Bedout structure, which is believed to be around 250 million years old, was considered a potential impact crater that could coincide with the timing of the Great Dying.

While some scientists have interpreted features in the area as consistent with an asteroid impact (such as shock-metamorphosed minerals), others argue that the geological features might be related to volcanic activity rather than an impact. Thus, the Bedout crater hypothesis remains controversial, and further evidence is needed to support the idea that an extraterrestrial event played a role in the Great Dying.

4.1.3 Limitations of the Impact Theory

The lack of strong evidence for an impact, such as a definitive crater or a global iridium layer, limits the support for this theory. Moreover, the duration of the extinction event, which took place over hundreds of thousands to millions of years, does not easily align with the idea of a sudden, catastrophic impact as the primary driver of extinction. While an impact may have contributed to environmental stress, it is unlikely to have been the sole cause of the Great Dying.

4.2 Marine Instabilities: The Role of Ocean Circulation and Anoxia

Another theory focuses on changes in **ocean circulation patterns** and their role in the **onset of anoxia**—the depletion of oxygen in the world's oceans. **Marine anoxia** has long been recognized as a major factor contributing to the Permian-Triassic extinction, particularly in the marine realm where approximately 96% of species were wiped out.

4.2.1 Ocean Circulation and Climate Change

Throughout Earth's history, ocean circulation patterns have been closely linked to climate. During the late Permian, the formation of **Pangaea**, along with increased volcanic activity, likely caused significant alterations in these circulation patterns. As global temperatures rose due to greenhouse gas emissions from the **Siberian Traps**, the oceans would have become warmer, affecting the balance of oxygen levels in the water.

In normal conditions, ocean currents help mix oxygenated surface waters with deeper, oxygen-poor waters. However, during periods of global warming, **ocean stratification** occurs, where warmer surface waters prevent this mixing. As a result, large areas of the ocean became stagnant, leading to the development of anoxic, or oxygen-deprived, zones. Marine life, particularly species that depended on well-oxygenated environments, faced mass die-offs.

4.2.2 Hydrogen Sulfide and Oceanic Poisoning

Another consequence of anoxic conditions in the ocean is the potential for the buildup of **hydrogen sulfide (H2S)**, a highly toxic gas produced by certain anaerobic bacteria that thrive in oxygen-deprived environments. This phenomenon, known as **euxinia**, has been suggested as a contributing factor to the Great Dying.

If vast regions of the ocean became euxinic, hydrogen sulfide would have accumulated not only in the water but also in the atmosphere. This toxic gas would have killed marine organisms in an already struggling environment and, if released into the atmosphere, could have severely impacted terrestrial life as well, leading to widespread ecosystem collapse on land. Furthermore, hydrogen sulfide may have contributed to the **depletion of the ozone layer**, increasing the amount of harmful ultraviolet (UV) radiation reaching Earth's surface and further stressing both marine and land-based species.

4.2.3 Evidence of Anoxic Oceans

Geological evidence supporting the marine instability theory includes widespread deposits of **black shales,** which are indicative of anoxic conditions. Black shales form when organic matter accumulates in oxygen-poor environments, suggesting that large swaths of the Permian oceans were suffering from severe oxygen depletion. Isotopic analysis of these deposits further supports the idea that carbon cycling was disrupted during the Permian-Triassic extinction, likely as a result of the collapse of marine ecosystems.

While marine anoxia alone cannot explain the full extent of the extinction event, it is likely that it played a significant role, particularly in the marine environments that were so heavily impacted.

4.3 Supercontinent Formation: Pangaea and Climate Disruption

The **formation of Pangaea,** the giant supercontinent that existed during the late Permian, is another factor that may have contributed to the environmental changes leading to the Great Dying. The assembly of Pangaea had profound effects on global climate, ocean circulation, and ecosystems.

4.3.1 Climate Changes Due to Pangaea

The consolidation of Earth's landmasses into a single supercontinent drastically altered the planet's climate systems. With much of the Earth's landmass located far from oceans, the interior regions of Pangaea became increasingly **arid** and **desert-like**. Large portions of the supercontinent experienced extreme seasonal fluctuations, with scorching summers and freezing winters. This **continental climate** would have posed significant challenges for terrestrial organisms, particularly those adapted to more stable, humid conditions.

Additionally, the formation of Pangaea may have altered **monsoonal patterns** and disrupted the flow of ocean currents. With fewer coastlines and larger landmasses, the oceans became less efficient at distributing heat around the planet, exacerbating temperature extremes. These climate shifts would have had cascading effects on ecosystems, potentially leading to **habitat loss** and the breakdown of ecological networks.

4.3.2 Oceanic Stagnation and Salinity Changes

Pangaea's formation also affected ocean circulation patterns, contributing to **stagnation** in the deep oceans. The reduction in coastline length and the closure of smaller seas may have limited the ability of ocean currents to circulate efficiently. This stagnation, in turn, could have exacerbated the onset of anoxia in the deep oceans, contributing to the widespread marine die-offs observed during the extinction event.

Furthermore, the breakup of smaller ocean basins and the concentration of marine life in fewer, larger oceans may have altered **salinity levels** and nutrient availability, further stressing marine organisms.

4.3.3 Ecological Impact of Habitat Loss

As Pangaea formed, large swaths of **coastal habitats**—which are among the most biologically productive ecosystems on Earth—disappeared. Coastal regions are typically nutrient-rich and provide essential environments for both marine and terrestrial species. The loss of these habitats would have been devastating, particularly for marine species dependent on shallow, productive waters. This loss of biodiversity hotspots could have been a significant contributing factor to the severity of the extinction.

Conclusion: Multiple Causes for a Global Catastrophe

While the **Siberian Traps** theory remains the most widely accepted explanation for the Great Dying, it is clear that no single cause can fully account for the magnitude of the extinction event. Instead, the Great Dying was likely the result of multiple, interconnected factors—**volcanism, climate change, oceanic anoxia, and the formation of**

Pangaea—that created a perfect storm of environmental crises.

The **asteroid impact theory** remains intriguing but lacks the definitive evidence found in other mass extinctions. Meanwhile, **marine instabilities** and **anoxic oceans** appear to have played a significant role, especially in the devastating loss of marine biodiversity. Finally, the assembly of **Pangaea** altered climate and ecosystems on a global scale, further contributing to the collapse of life on Earth.

As research continues, scientists may uncover more clues about the intricate web of causes behind the Great Dying. What remains certain is that this mass extinction reshaped the course of life on Earth, paving the way for the rise of the dinosaurs in the Triassic period.

Chapter 5: Lessons from the Great Dying

The **Great Dying**, or the **Permian-Triassic mass extinction**, offers modern science an invaluable glimpse into the delicate balance that governs Earth's ecosystems. As the most catastrophic extinction event in the history of life, it serves as a powerful reminder of the fragility of our planet's environment and the complex interactions between climate, atmosphere, and biodiversity. While the event occurred 252 million years ago, the insights gleaned from studying it have profound implications for our understanding of modern challenges, particularly **climate change**, **mass extinction events**, and the **resilience of life** in the face of extreme environmental disruptions.

5.1 Climate Change: A Cautionary Tale

The dramatic shifts in climate during the Permian-Triassic extinction provide a historical precedent for understanding the consequences of rapid **greenhouse gas emissions**. One of the primary drivers of the Great Dying was the massive release of **carbon dioxide (CO_2)** into the atmosphere due to the **Siberian Traps** volcanic eruptions. This led to a cascade of environmental catastrophes, including **global warming**, **ocean acidification**, and **anoxia** (the depletion of oxygen in oceans), which proved fatal for a vast majority of Earth's species. The parallels between the climate shifts during the Great Dying and modern **anthropogenic climate change** are stark, offering critical lessons:

5.1.1 Greenhouse Gas Emissions and Global Warming

During the Great Dying, the release of CO_2 into the atmosphere from the Siberian Traps likely caused global temperatures to increase by as much as **10-15°C**. This extreme warming had cascading effects on ecosystems worldwide. Modern climate change, driven by the burning of **fossil fuels** and the release of CO_2 and **methane** into the atmosphere, mirrors this ancient event on a smaller, yet significant scale.

The Permian-Triassic extinction demonstrates the potentially devastating impact of unchecked greenhouse gas emissions. The dramatic temperature rise contributed to the **acidification of oceans**, which in turn led to the collapse of marine ecosystems, as many species were unable to adapt to the changing chemistry of their environment. The Great Dying warns that if CO_2 levels continue to rise at their current rate, modern marine ecosystems, including **coral reefs** and **shellfish populations**, may face a similar fate.

5.1.2 Ocean Acidification and Anoxia

One of the most lethal consequences of the Great Dying was the **acidification of the oceans**. As CO_2 levels rose, much of the gas dissolved into the oceans, forming **carbonic acid**. This process reduced the pH of seawater, making it difficult for marine organisms, such as **corals**, **mollusks**, and **plankton**, to build their calcium carbonate skeletons and shells. The collapse of these foundational species triggered a domino effect, leading to the extinction of entire marine food chains.

Furthermore, warming temperatures caused the oceans to lose oxygen, a condition known as **oceanic anoxia**. Oxygen-depleted waters are hostile to most marine life, and during the Permian-Triassic event, vast regions of the world's oceans became "dead zones." This scenario offers a clear warning to modern scientists and policymakers: as global temperatures rise due to human activities, the risk of ocean acidification and anoxia increases, with potentially catastrophic consequences for biodiversity and food security.

5.1.3 Feedback Loops and Tipping Points

The Great Dying also illustrates the dangers of **feedback loops** and **climatic tipping points**. As global temperatures rose, they likely triggered the release of methane, a potent greenhouse gas, from undersea methane hydrates. This sudden release of methane would have further accelerated warming, creating a feedback loop that made the climate crisis even more severe. Once certain thresholds were crossed, the climate spiraled out of control, leading to a runaway greenhouse effect that Earth could not recover from quickly.

In the modern era, scientists are concerned about similar feedback loops, such as the melting of **permafrost** in the Arctic, which could release vast amounts of methane into the atmosphere, intensifying global warming. The lesson from the Great Dying is that once these tipping points are crossed, it may become impossible to reverse the damage, emphasizing the urgency of addressing climate change before it reaches a point of no return.

5.2 Mass Extinction Events: Predicting Future Crises

The Great Dying was not just a climatic event; it was also the most devastating **mass extinction** in Earth's history, eliminating an estimated **90-96% of marine species** and **70% of terrestrial species**. By studying the causes and consequences of this event, scientists can gain a better understanding of how future mass extinctions might unfold and how species respond to environmental stresses.

5.2.1 Biodiversity Loss and Ecosystem Collapse

One of the key lessons from the Permian-Triassic extinction is the vulnerability of **biodiversity** to rapid environmental change. As ecosystems become more stressed—whether by climate change, habitat loss, or pollution—the ability of species to adapt diminishes. The Great Dying saw the collapse of complex ecosystems, with highly specialized species being particularly hard hit. In contrast, more **generalist species** that could survive in a range of conditions were more likely to endure.

In today's world, human activities are driving species toward extinction at an alarming rate. According to recent studies, Earth is currently undergoing its **sixth mass extinction**, with species disappearing at **100 to 1,000 times** the natural background rate. Habitat destruction, overexploitation, pollution, and climate change are the primary culprits. The lesson from the Great Dying is that once ecosystems begin to collapse, the process can accelerate rapidly, with cascading effects across the globe. Protecting biodiversity today is critical to preventing a repeat of such a devastating event.

5.2.2 Species Adaptation and Survival

The Permian-Triassic extinction also offers insights into how some species manage to survive mass extinction events. During the Great Dying, certain groups of organisms, such as **cyanobacteria** and **bivalves**, managed to survive and even thrive in the harsh, anoxic conditions of the oceans. On land, small, **burrowing animals** were more likely to survive because they could escape the extreme heat and arid conditions on the surface.

This suggests that species with **adaptive traits**, such as the ability to live in a wide range of environments, **resilience to changing conditions**, or **rapid reproductive cycles**, are more likely to survive future mass extinction events. By understanding the traits that allowed certain species to endure the Great Dying, scientists can predict which modern species might be more resilient to current environmental changes and which are at greater risk.

5.2.3 Cascading Effects and Global Impacts

The Great Dying also highlights the interconnectedness of ecosystems. When one group of organisms—such as **coral reefs** or **forest ecosystems**—collapses, it can trigger a chain reaction that affects other species and habitats. During the Permian-Triassic extinction, the collapse of marine ecosystems had ripple effects across terrestrial environments, as the loss of marine biodiversity impacted global nutrient cycles and climate systems.

In the modern world, we are already witnessing similar cascading effects. The decline of pollinator populations, for example, threatens global food security, while the loss of coral reefs reduces the biodiversity of marine ecosystems and weakens coastal protection from storms. The lesson from the Great Dying is clear: the health of one ecosystem is linked to the health of all ecosystems, and the loss of key species or habitats can have far-reaching consequences.

5.3 Resilience of Life: Rebirth After Devastation

Despite the overwhelming destruction caused by the Great Dying, life on Earth eventually recovered. This recovery, however, took **millions of years** and was marked by the emergence of entirely new species and ecosystems. The resilience of life in the face of such devastation provides a hopeful message about the adaptability and endurance of life on Earth, even in the most extreme conditions.

5.3.1 Evolutionary Innovation and Adaptation

The Permian-Triassic extinction event cleared the way for a **wave of evolutionary innovation**. With many ecological niches left vacant by the extinction, new groups of organisms emerged to fill them. Among the most notable were the **archosaurs**, the ancestors of **dinosaurs**, which rose to dominance in the Triassic period. **Mammal-like reptiles**, which had been a dominant group in the Permian, were largely replaced by these new species, showcasing how extinction events can dramatically shift the course of evolution.

This pattern of destruction followed by innovation has been observed after every mass extinction event in Earth's history. While the loss of biodiversity is tragic, the Great

Dying reminds us that life is incredibly resilient and capable of adapting to new conditions. In the face of modern environmental crises, this adaptability offers hope that, with proper conservation efforts, life on Earth can continue to thrive even in a changing world.

5.3.2 The Long Road to Recovery

While life eventually rebounded after the Great Dying, the recovery process was slow and arduous. It took **millions of years** for biodiversity levels to approach pre-extinction levels, and the ecosystems that emerged were often drastically different from those that existed before the event. This slow recovery emphasizes the importance of **preventative measures** in the modern era. While life may be resilient, the restoration of ecosystems and biodiversity can take an exceedingly long time. The lesson here is that we cannot afford to rely on nature's ability to "bounce back" after environmental disasters—proactive efforts to protect ecosystems and prevent mass extinctions are essential.

5.3.3 Lessons for the Future

The resilience of life seen in the aftermath of the Great Dying serves as both a cautionary tale and a source of inspiration. While the recovery of life after the extinction demonstrates the incredible adaptability of species, the magnitude of loss and the time it took for life to rebound

The Great Dying

underscore the need for urgent action in the face of current environmental threats.

Conclusion: A Window into Earth's Past

The **Great Dying**, also known as the **Permian-Triassic extinction event**, stands as a stark reminder of the fragility of life on Earth and the interconnectedness of ecosystems. This catastrophic event, which wiped out nearly all forms of life around **252 million years ago**, reshaped the planet in ways that are still relevant to our understanding of Earth's history and future. It serves as a profound example of how environmental changes—whether sudden or gradual—can have devastating consequences for life on a global scale.

A Catastrophic Turning Point

The magnitude of the Great Dying cannot be overstated. It marks the single most severe mass extinction event in Earth's history, with nearly **96% of all marine species** and **70% of terrestrial vertebrates** wiped out in a relatively short geological period. This extinction event effectively ended the **Permian period** and heralded the start of the **Triassic**, a new era in which Earth's ecosystems had to rebuild from near total devastation.

The causes of the Great Dying—while still debated in some respects—are generally attributed to **massive**

volcanic eruptions, particularly from the **Siberian Traps**, which released enormous amounts of greenhouse gases into the atmosphere. These eruptions triggered a cascade of environmental disasters: **global warming, ocean acidification**, and **widespread anoxia** in the oceans, where oxygen levels plummeted. The resultant chain of climate events led to the collapse of ecosystems on land and sea alike.

What makes the Great Dying so significant is not just the scale of destruction, but also the breadth of its impact across multiple environments. From the coral reefs that once thrived in the oceans to the diverse synapsids that roamed the land, nearly every ecosystem was decimated. Entire groups of animals, such as trilobites, which had been around for hundreds of millions of years, were completely wiped out. The marine food web collapsed, and even the hardy terrestrial species faced severe challenges in the aftermath.

From Destruction to Rebirth

Yet, despite the overwhelming loss of life, the Great Dying was not the end of life on Earth. In fact, this mass extinction, while catastrophic, also set the stage for the next great chapter in Earth's history—the rise of the **dinosaurs** and, eventually, **mammals**. Life slowly began to rebuild itself during the **Triassic period**, as new species evolved to fill the ecological niches left vacant by those that perished.

This post-extinction recovery took millions of years, as ecosystems gradually stabilized. **Archosaurs**, a group that would eventually give rise to both dinosaurs and crocodiles, began to dominate the land. In the oceans, new forms of fish and marine reptiles emerged, adapting to the changing environment. Although it took millions of years for life to regain the level of diversity seen before the Great Dying, this period of renewal is a testament to the **resilience of life** on Earth.

The extinction event also paved the way for a reconfiguration of Earth's evolutionary pathways. Had the Permian extinction not occurred, it is possible that **mammals** as we know them today would never have evolved, as they emerged from survivors of the event, such as small synapsids. The **Age of Reptiles**, or **Mesozoic Era**, would follow, leading to the dominance of the dinosaurs for the next 160 million years.

Lessons from Earth's Ancient Catastrophe

The Great Dying holds profound lessons for us in the modern era. While the causes of this ancient extinction were primarily natural, involving massive volcanic eruptions and climatic shifts, it highlights how vulnerable life is to rapid environmental changes. In our current geological epoch, the **Anthropocene**, human activities are accelerating climate change at an unprecedented rate. Rising **carbon dioxide levels, ocean acidification**, and **biodiversity loss** are challenges eerily similar to those faced during the Permian-Triassic extinction.

By studying the Great Dying, scientists have gained invaluable insights into how ecosystems respond to severe environmental stressors. The knowledge of how certain species survived, adapted, or evolved during the aftermath of the Great Dying provides clues about the **resilience** and **fragility** of life in the face of mass extinctions. This understanding is crucial as we face ongoing threats from **climate change, habitat destruction**, and other anthropogenic impacts on the biosphere.

The **volcanic eruptions** that caused the Great Dying released immense amounts of **greenhouse gases**, triggering runaway global warming and oceanic disruptions. Today, the burning of fossil fuels is producing similar gases, and while the timescales differ, the potential for widespread environmental damage is alarmingly similar. The possibility of tipping points—where climate systems can shift irreversibly—is something scientists are closely monitoring in light of past events like the Great Dying.

Moreover, the extinction also offers a warning about the **delicate balance of ecosystems**. In the Permian, the complex web of life that supported vast marine and terrestrial ecosystems collapsed, leaving the planet nearly lifeless. Today, we are witnessing declines in biodiversity at rates that many experts consider to be the onset of a **sixth mass extinction**. Protecting ecosystems and preventing biodiversity loss are essential steps in ensuring that our planet remains habitable for future generations.

Hope and the Future of Life on Earth

While the Great Dying was a period of immense loss, it also offers hope. Life on Earth, despite its vulnerability, is remarkably resilient. Even in the face of near-total extinction, life found a way to bounce back, evolve, and flourish once again. This resilience is a key takeaway for us today as we grapple with the consequences of environmental degradation and climate change.

By understanding the forces that caused the Great Dying, we can better prepare for and mitigate the challenges ahead. We can also learn from the survivors of that ancient catastrophe—the species that adapted to new conditions and those that filled the niches left behind. Their evolutionary success offers a blueprint for how life can endure, adapt, and thrive in the face of extreme adversity.

In conclusion, the **Great Dying** is not just a story of destruction—it is also a story of **renewal**. As we look back at this pivotal event in Earth's history, we are reminded that life, despite being battered by unimaginable forces, is tenacious and adaptable. The lessons learned from this mass extinction provide a window into Earth's deep past, while also serving as a guide for navigating the uncertain environmental future we face today. The **Great Dying** teaches us that while the stakes are high, life on Earth has the potential to endure and recover, just as it did 252 million years ago. However, the choices we make today about how we interact with our environment will determine whether that recovery includes humanity.

Glossary

Anoxia
A condition in which there is a severe depletion or total absence of oxygen, especially in marine environments. Anoxia was a key factor during the Great Dying, leading to the collapse of ocean ecosystems.

Archosaurs
A group of diapsid reptiles that includes modern birds and crocodiles, as well as extinct dinosaurs and pterosaurs. After the Great Dying, archosaurs became dominant during the Triassic period.

Biodiversity
The variety and variability of life on Earth, encompassing different species, ecosystems, and genetic diversity. The Great Dying caused one of the largest reductions in biodiversity in Earth's history.

Brachiopods
Marine organisms with hard "valves" (shells) on their upper and lower surfaces, resembling clams. They were abundant in the oceans before the Great Dying but were greatly reduced during the extinction.

Carbon Dioxide (CO2)
A greenhouse gas produced by natural processes such as

volcanic eruptions and human activities. Massive volcanic emissions of CO2 during the Permian period are thought to have contributed to global warming and the environmental changes during the Great Dying.

Clathrate (Methane Hydrate)

A solid ice-like compound in which methane molecules are trapped within a structure of water molecules. Methane hydrates, located in ocean sediments, may have been released as global temperatures increased during the Great Dying, contributing to further warming.

Coral Reefs

Large underwater structures made from calcium carbonate secreted by corals. Reefs are important ecosystems for marine life. The coral reefs of the Permian period were destroyed by ocean acidification and temperature changes during the Great Dying.

Diapsid

A major group of reptiles that includes most modern reptiles, such as lizards, snakes, and crocodiles, as well as extinct groups like dinosaurs. Diapsids diversified after the Great Dying.

Extinction

The complete disappearance of a species from Earth. The Permian-Triassic extinction event, also known as the Great

Dying, saw the extinction of a vast number of species across marine and terrestrial ecosystems.

Greenhouse Effect

The warming of Earth's surface due to the trapping of heat by certain gases in the atmosphere, such as carbon dioxide and methane. The extreme greenhouse effect during the Great Dying led to rapid global temperature increases.

Methane

A potent greenhouse gas that can be released from natural sources like wetlands, undersea deposits, and during volcanic activity. During the Great Dying, the release of methane from ocean sediments exacerbated global warming.

Pangaea

The supercontinent that existed during the late Paleozoic and early Mesozoic eras, comprising almost all the landmasses on Earth. The formation of Pangaea is thought to have contributed to climate changes during the Great Dying by altering ocean currents and weather patterns.

Permian Period

The last geological period of the Paleozoic Era, lasting from approximately 298 to 252 million years ago. The Permian ended with the most significant mass extinction event, the Great Dying.

Siberian Traps
A vast region of volcanic rock in modern-day Russia, which formed due to massive volcanic eruptions during the late Permian period. These eruptions released enormous amounts of volcanic gases, contributing to the environmental crises that caused the Great Dying.

Synapsids
A group of animals that includes mammals and their extinct relatives. Synapsids were the dominant land animals during the Permian period but were heavily impacted by the Great Dying, leading to the rise of archosaurs and dinosaurs.

Trilobites
An extinct group of marine arthropods that thrived for hundreds of millions of years before the Permian-Triassic extinction. Trilobites were one of the many marine species that disappeared during the Great Dying.

Volcanic Outgassing
The release of gases such as carbon dioxide and sulfur dioxide during volcanic eruptions. The massive volcanic outgassing from the Siberian Traps is believed to have played a significant role in the environmental changes during the Great Dying.

Zooplankton
Small and often microscopic animals that float or drift in

water, serving as a critical part of the marine food chain. Zooplankton populations were severely affected by ocean anoxia and acidification during the Great Dying.

Bibliography

1. Extinction: How Life on Earth Nearly Ended 250 Million Years Ago by Douglas H. Erwin

This authoritative work delves deep into the Permian-Triassic extinction, offering detailed scientific explanations of the environmental causes and biological impacts.

2. When Life Nearly Died: The Greatest Mass Extinction of All Time by Michael J. Benton

A comprehensive look at the events that unfolded during the Great Dying, this book provides insights into the geological and paleontological evidence behind the extinction.

3. Under a Green Sky: Global Warming, the Mass Extinctions of the Past, and What They Can Tell Us About Our Future by Peter D. Ward

Ward connects past extinction events, including the Great Dying, to modern concerns about climate change, offering a relevant perspective for today's environmental issues.

4. The Ends of the World: Volcanic Apocalypses, Lethal Oceans, and Our Quest to Understand Earth's Past Mass Extinctions by Peter Brannen

This book offers an engaging narrative on various mass extinctions, with a particular focus on the events and processes behind them, including the Permian-Triassic event.

5. **The Sixth Extinction: An Unnatural History** by Elizabeth Kolbert

While focusing on the current mass extinction, Kolbert's work provides a framework for understanding extinction events, helping to draw parallels between the Great Dying and the present biodiversity crisis.

6. **Mass Extinctions and Their Aftermath** by A. Hallam and P.B. Wignall

A detailed examination of mass extinctions throughout Earth's history, this academic book focuses on the causes and consequences, with substantial coverage of the Permian extinction.

7. **Catastrophes and Lesser Calamities: The Causes of Mass Extinctions** by Anthony Hallam

Hallam's book offers a concise but informative look at various mass extinctions, including the Permian-Triassic event, with an emphasis on the geological and climatological processes involved.

8. **The Permian Extinction and the Tethys: An Exercise in Global Geology** by David H. K. Georis

This book offers a geological perspective on the Permian extinction, examining the role of tectonics and oceanic changes, specifically the development of the Tethys Ocean.

Acknowledgments

Writing *The Great Dying: Earth's Most Catastrophic Mass Extinction* has been an immensely rewarding journey, and I owe a great deal of gratitude to the many individuals who have supported and inspired me along the way.

First and foremost, I would like to express my heartfelt thanks to my family and friends for their unwavering support and encouragement. Their belief in my work has been a constant source of motivation, and I am deeply grateful for their patience as I immersed myself in the world of ancient extinctions and ecological upheavals.

I would like to extend my gratitude to the many scientists, paleontologists, and researchers whose groundbreaking work has paved the way for a deeper understanding of the Permian-Triassic extinction. Their tireless efforts to uncover the mysteries of Earth's past have been instrumental in shaping this book. Special thanks to the work of **Douglas H. Erwin**, **Michael J. Benton**, and **Peter D. Ward**, whose books and research were invaluable resources throughout the writing process.

A sincere thank you goes to my editor, whose keen eye and insightful feedback helped refine this book into its final form. Your expertise has greatly enhanced the clarity and readability of this work.

To my peers and colleagues in the field of science communication, thank you for your input and discussions, which helped shape the narrative of this book. Your passion for making science accessible to all has been truly inspiring.

Lastly, I am grateful to the readers, past and present, who share my fascination with Earth's ancient history and the incredible events that have shaped life as we know it today. It is my hope that this book sparks the same curiosity and wonder that I have felt while studying the Great Dying and its far-reaching implications.

Thank you all for being part of this journey with me.

Sincerely,

Zahid Ameer
Versatile Indie Author

Disclaimer

The information presented in this book, **'The Great Dying: Earth's Most Catastrophic Mass Extinction,'** is based on current scientific research, historical data, and the author's interpretation of available resources at the time of writing. While every effort has been made to ensure accuracy, the fields of paleontology, geology, and climate science are constantly evolving, and new discoveries may lead to revisions in our understanding of past events.

This book is intended for educational and informational purposes only and should not be considered a definitive or exhaustive guide. The author and publisher assume no responsibility for any inaccuracies, errors, or omissions that may arise, nor for any actions taken based on the content of this book. Readers are encouraged to consult additional sources and engage with current scientific literature for a more comprehensive understanding of the subject matter.

All opinions expressed in this book are those of the author and do not necessarily reflect the views of scientific institutions or organizations.

About me

I am Zahid Ameer, hailing from the vibrant country of India. As an author, ghostwriter, bibliophile, online affiliate marketer, blogger, YouTuber, graphic designer, and animal lover, I have woven my passions into a unique tapestry that defines my life's work.

Born and raised in India, I have always possessed a deep love for literature. With an insatiable appetite for books, I have amassed an impressive collection of around 1,600 titles, predominantly in English. My passion for reading brings me immense joy and serves as a source of inspiration for my writing endeavors.

I have compiled an impressive portfolio of written works as an author and ghostwriter. With a captivating writing style and an innate ability to craft engaging narratives, I bring my stories to life, captivating readers from all walks of life. My wide range of interests and experiences contribute to the richness of my writing, allowing me to connect with my audience on a heartfelt level effortlessly.

Beyond my literary pursuits, I have also established a strong presence on various digital platforms. I utilize my YouTube channel and blog to raise awareness about all types of knowledge and to share heartwarming stories of animals. Using my platform to shed light on important issues, I strive to create a world where humans and animals can coexist harmoniously.

In addition to my work as an author, I have also dabbled in the world of affiliate marketing. With my webpreneur spirit, I have ventured into online marketing, leveraging my knowledge and skills to promote products and services that align with my values.

However, my most cherished role is that of a father. Family is at the core of my being, and everything I do is centered around creating a better future for my loved ones. My dedication to my family is evident in my passion for personal growth and my relentless pursuit of success. Through my various endeavors, I strive to set an example of perseverance and ambition for my children, inspiring them to chase their dreams unapologetically.

In a world where specialization often dominates, I defy convention by embracing multiple passions and excelling in diverse fields. My love for books, animals, and family has become the driving force behind my achievements. By the grace of Almighty God, my unique blend of characteristics has allowed me to leave an indelible mark on the world, enriching the lives of those I encounter along the way.

To your grand success in life,

Zahid Ameer
Versatile Indie Author